Petroleum Geology for Geophysicists and Engineers

Richard C. Selley

International Human Resources Development Corporation • Boston

ISBN: 0-934634-49-1   Cloth
       0-934634-42-4   Paper

Library of Congress Catalog Card Number: 82-81124

Printed in the United States of America

Contents

Author's Note   *vii*

1 **Hydrocarbons**   1

   Introduction   1
   The Physical and Chemical Properties of
     Hydrocarbons   1
   Five Parameters Necessary for Hydrocarbon
     Accumulation   3
   Mode of Distribution of Oil and Gas   3

2 **Hydrocarbon Source Rocks**   7

   Introduction   7
   Generation of Organic Matter and Preservation
     in Clays   8
   Diagenesis and Maturation of Organic Matter   9
   Migration of Hydrocarbons   11

3 **Geophysical Well Logs**   15

   Introduction   15
   Geophysical Logging Introduction   15
   The Spontaneous Potential Log   17
   The Resistivity Log   17
   The Gamma Log   18
   The Neutron Log   18

The Density Log 18
The Sonic Log 19
The Dipmeter Log 21
Geophysical Logging Summary 23
Integrated Facies Analysis 23

4 Reservoir Rocks I 31

Introduction: Porosity and Permeability 31
Sandstone Reservoirs 34
Depositional Processes and Environments 35

5 Reservoir Rocks II: Carbonates 41

Introduction 41
Carbonate Mineralogy and Components 42
Carbonate Environments and Facies 43
Reefs 45

6 Evaporites 51

Introduction 51
Occurrence and Formation of Evaporites 52
Significance of Evaporites to Petroleum
    Exploration 54

7 Hydrocarbon Traps 57

Introduction 57
Basic Parameters of a Trap 57
Classification of Hydrocarbon Traps 57
Structural Traps 58
Hydrodynamic Traps 59

8 Stratigraphic Traps 63

Introduction 63
Stratigraphic Trap Exploration Techniques 63
Unconformity-Associated Traps 64

Stratigraphic Traps Unassociated with
Unconformities   65

9   The Habitat of Hydrocarbons in Sedimentary Basins   69

Introduction   69
Mechanisms of Basin Genesis   69
Distribution of Hydrocarbons within a
Sedimentary Basin   75

Appendix: Review of the Petroleum Geology of the
North Sea   77
Introduction   77
The Geological History of the North Sea   77
The Habitat of Hydrocarbons in the North Sea   81

References   *85*

Index   *87*

Author's Note

This book has been designed to concisely explain the geological concepts which govern petroleum exploration, to summarize the data on which these concepts are based, and to review some of the geological techniques used in petroleum exploration, e.g., geochemical analysis and geophysical well logging.

It has been written for non-geologists in the oil industry, primarily geophysicists and petroleum engineers. The book assumes a familiarity with basic geological concepts and terminology. Readers who lack this background may find it useful to keep a geological dictionary at hand.

1

Hydrocarbons

Introduction

The first chapter begins by describing the physical and chemical properties of hydrocarbons and defining five parameters that must be fulfilled for a commercial accumulation of hydrocarbons to occur. These *magic* five will then be examined in greater detail throughout the book.

The chapter concludes by reviewing the modal distribution of hydrocarbons, with respect to age, trap type, reservoir rock type, basinal setting and depth, and by discussing reasons for this pattern.

The Physical and Chemical Properties of Hydrocarbons

There are five major types of hydrocarbons of interest to petroleum exploration.

Kerogen. Kerogen is fine-grained, generally amorphous, organic matter found principally in argillaceous sediments. Kerogen generates crude oil when heated and is insoluble in normal petroleum solvents such as carbon disulphide. Its average chemical composition is 75% carbon, 10% hydrogen, and 15% other (sulfur, oxygen, nitrogen, etc.).

Kerogen is of no commercial significance except where it is so abundant (greater than 10%) as to occur in oil shales. It is, however, of great geological importance because it is the substance which generates hydrocarbon oil and gas. A *source rock* must contain significant amounts of kerogen.

Asphalt. Asphalt is a hydrocarbon that is solid at surface temperatures and pressures, though it may flow slowly and plastically. Unlike kerogen, asphalt is soluble in normal petroleum solvents. It is produced by the partial maturation of kerogen or by the degradation of mature crude oil.

Crude Oil. Crude oil is a hydrocarbon that is liquid at surface temperatures and pressures, and is soluble in normal petroleum solvents. Chemically, it consists of four major groups of organic compounds: paraffins, naphthenes, aromatics, and resins (heterocompounds containing sulphur, nitrogen, oxygen, and various odd metals).

Crude oil may be classified chemically (e.g., paraffinic, naphthenic, etc.) or by its density. This may be expressed as specific gravity or as API gravity according to the formula:

$$\text{API}° = \frac{141.5}{\text{sp. grav. @ } 60°\text{F}} - 131.5. \qquad (1)$$

An API gravity of 10 equals a specific gravity of 1.0. Heavy oils, e.g., those that sink in water, have an API of less than ten. Normal crudes have API values of between 30-40° API (0.88 − 0.80 specific gravity).

Natural Gas. Natural gas is a term used in the industry to imply hydrocarbon gas, though there are many inorganic natural gases beneath the earth. The major hydrocarbon gases are: methane (CH_4), ethane (C_2H_6), propane (C_3H_8), and butane (C_4H_{10}). Hydrogen sulphide (H_2S) gas is sometimes present, hence the terms *sour* and *sweet* gas.

Condensates. Condensates are hydrocarbons transitional between gas and crude oil (gaseous in the subsurface but condensing to liquid at surface temperatures and pressures). Chemically, condensates consist largely of paraffins, such as pentane, octane, and hexane.

Five Parameters Necessary
for Hydrocarbon Accumulation

It is a matter of observation that commercial quantities of oil and gas occur only in sedimentary basins. Five conditions are necessary for their entrapment: (1.) A source rock is required to generate hydrocarbons. This is generally a fine-grained clay containing over 0.5% organic matter (kerogen) by weight. (2.) A reservoir rock is required to store hydrocarbons. It must be porous to hold the oil or gas and permeable to allow for its production. (3.) A seal, or cap rock, which is impermeable so as to prevent the upward escape of hydrocarbons from the reservoir. (4.) A trap occurs when the source, reservoir, and seal are arranged in such a geometry that hydrocarbons may migrate from the source into a sealed reservoir wherein they are trapped. (5.) The source rock must have been heated sufficiently to generate oil (greater than 60 °C) or gas (greater than 150 °C). Regions barren of hydrocarbons occur where there has been overheating (greater than 250 °C) or where the source rock is still immature.

These five parameters will be examined in detail in the next few chapters.

Mode of Distribution of Oil and Gas

Data compiled by Moody et al. (1970) show how giant oil and gas fields occur (a giant field contains 500 million barrels (mb) of recoverable oil or 3.5 trillion cubic feet (tcf) of recoverable gas). Note that these data refer only to discovered fields, not necessarily those that are actually there.

The data in tables 1–5 lead to the conclusion that, in order to find a major oil field, a shelf where there is an anticline with Cretaceous sandstone at about 8000 feet should exist. It is interesting to speculate why this is so.

TABLE 1
DISTRIBUTION OF OIL AND GAS FIELDS BY GEOLOGIC AGE

Geologic Age	% of Fields
Neogene	18
Palaeogene	21
Cretaceous	27*
Jurassic	21
Permo-Trias	6
Carboniferous	5
Devonian	1
Cambrian-Silurian	1
	100

*major class

TABLE 2
DISTRIBUTION OF OIL AND GAS FIELDS BY DEPTH OF FIELD

Depth of Field (ft in thousands)	% of Fields
0-2	2
2-4	8
4-6	24
6-8	42*
8-10	12
10-12	10
12-14	2
14-	0
	100

*major class

TABLE 3
DISTRIBUTION OF OIL AND GAS FIELDS BY TYPE OF TRAP

Type of Trap	% of Fields
Structural	
Anticline	75*
Fault	1
Salt Dome	2
	78
Stratigraphic	
Reef	6
Pinchout	5
Truncation	2
	13
Combination	9
	9
	100

*major class

TABLE 4
DISTRIBUTION OF OIL AND GAS FIELDS
BY POSITION IN BASIN OF FIELDS

Position in Basin	% of Fields
Shelf	70*
Basin Center	10
Mobile Trough (Orogen)	20
	100

*major class

TABLE 5
DISTRIBUTION OF OIL AND GAS FIELDS
BY LITHOLOGY OF RESERVOIR

Lithology of Reservoir	% of Fields
Sandstone	49*
Carbonate	45
Igneous/Metamorphic	6
	100

*major class

2

Hydrocarbon Source Rocks

Introduction

This chapter examines the evidence for the organic origin of oil and gas, shows how source rocks may be identified, how hydrocarbons develop from kerogen, and how hydrocarbons migrate from source to reservoir. It therefore shows how geologists seek to answer three basic questions in an exploration province.

(1.) Which strata, if any, are hydrocarbon source rocks?
(2.) Is the source material oil prone or gas prone?
(3.) What is the degree of maturation of the source rock? Is it immature (too cool to have generated), mature (just right), or overmature (overheated and barren)?

There is much evidence for the abiogenic formation of organic compounds in outer space (e.g., carbonaceous chondrites) and in igneous rocks (e.g., the intrusives of the Khibiny massif in the U.S.S.R.). There is abundant evidence, however, that commercial quantities of oil and gas are generated in sedimentary rocks at relatively low (submetamorphic) crustal temperatures. Most geologists believe that these hydrocarbons are produced by the thermal maturation of organic matter buried synchronously with the sediments. Critical evidence includes the following facts.

7

Crude oil exhibits the property of laevorotation (the property of rotating polarized light), which is found only in biosynthesized organic compounds. Crude oil contains complex organic molecules, such as porphyrins and steroids, found elsewhere only in organic tissues and fluids. Oil frequently occurs trapped in porous, permeable, reservoir sediments totally enclosed in impermeable shales (generally rich in kerogen). Where commercial quantities of oil and gas occur in igneous or metamorphic rocks, they generally intrude or are uncomformably overlain by organic-rich shales.

Generation of Organic Matter and Preservation in Clays

In the ordinary course of events, organic matter on the surface of the earth is destroyed by the normal processes of oxidation, fermentation, and bacterial decay. The two factors that control the preservation of organic matter are the *rate of generation* and the *rate of destruction*.

Obviously, generation must be greater than destruction for organic matter to be buried. Factors that enhance organic generation on land include high temperature, abundant moisture, and long daylight hours. In marine environments, organic productivity is greatest at shallow depths, where warm water mixes with cool, nutrient-laden water from the deep ocean. These conditions are met on the west side of the continents and in intraoceanic rifts.

A factor that enhances the preservation of organic matter on land is waterlogged soil (as in swamps). Factors that enhance the preservation of organic matter underwater include rapid sedimentation and thermal layering. Organic matter tends to be preserved in low permeability, fine-grained sediments, rather than permeable sands in which flushing by oxygenated ground water may destroy organic matter.

It is worth looking at the mineralogy of clays since they are related to hydrocarbon preservation and generation. There are three major clay minerals.

Kaolin ($Al_2O_3 2SiO_2 \bullet 2H_2O$). Kaolin forms in low pH (acid) environments. At the earth's surface these are, generally, also oxidizing. Thus kaolin clays generally contain little organic matter and characterize continental environments.

Illite [K Al$_2$(OH)$_2$AlSi$_3$(O,OH)$_{10}$]. Illite forms in alkaline-reducing conditions that are characteristic of marine environments. Because of their affinity for reducing conditions, illitic clays often preserve organic matter.

Montmorillonite (smectite or bentonite) [(MgCa)O•Al$_2$O$_3$ 5SiO$_2$•nH$_2$O]. Montmorillonite, like illite, prefers reducing environments and, therefore, may also be associated with organic matter. Sometimes, it is produced by the devitrification of volcanic glass. Montmorillonite contains abundant water molecules between the lattice layers. Expulsion of this water during burial may aid the emigration of hydrocarbons. These are the dreaded *swelling* clays.

Diagenesis and Maturation of Organic Matter

When preserved organic matter is buried, it undergoes three phases of alteration.

Diagenesis. Diagenesis occurs at normal temperatures and pressures within the top few meters of the sediment column. Organic matter may be subjected to bacterial decay, oxidation, dehydration, and decarboxylation. Thus H$_2$O, CO$_2$, and biogenic methane (CH$_4$) may be expelled. The result is kerogen. Clay compaction, at the same time, rapidly reduces porosity from some 60% to 40%.

Catagenesis. Catagenesis occurs as temperatures increase to about 250 °C. Kerogen generates oil and gas. Shale porosity decreases to $\simeq$ 10%.

Metagenesis. Metagenesis occurs above 250 °C. On the threshold of metamorphism, kerogen consists now of little more than carbon (graphite).

Catagenesis is obviously of critical importance and must be examined in more detail. The first major factor to consider in catagenesis is the nature of the kerogen. Three major types of kerogen can be recognized.

Type I Kerogen (Algal). Chemically, type I kerogen is very rich in hydrogen, relatively low in oxygen, and contains lipids, oils, fats, and waxes. These are particularly abundant in algae, both marine and freshwater. Type I kerogen tends to generate oil and is present in may oil shales.

Type II Kerogen (Liptinic). Rich in aliphatic compounds, with a hydrogen-carbon (H:C) ratio greater than one, type II kerogen originally formed from algal detritus and zooplankton and phytoplankton. The Kimmeridge clay of the North Sea is of this type. Type II kerogen can generate both oil and gas.

Type III Kerogen (Humic). Type III kerogen has a low H:C ratio (less than 0.84), is low in aliphatics and rich in aromatic compounds. Humic kerogen is produced from the lignin of higher woody plants. This type of kerogen is gas prone.

The chemistry of these three types of kerogen is shown in Figure 1.

Type I and II kerogens tend to occur in marine environments, type III in continental (fluvial and deltaic) ones. Hence the broad generalization made by geologists that marine source rocks tend to generate oil, but continental ones tend to yield gas.

The second major factor to consider in catagenesis is temperature. It is now generally agreed that the three kerogen types mature and generate hydrocarbons as temperature increases. Major oil generation commences at about 60°C, reaches an optimum at about 120°C, and thereafter declines, to be replaced by gas generation. Oil generation totally dies out at about 150°C and gas generation at about 250°C. It is therefore important to know the thermal maturity of a source rock.

Temperature determined from a borehole only gives present-day value. We need palaeothermometers which can measure the maximum temperature to which a source rock has ever been heated. Current methods used include: the ratio of total organic carbon to residual carbon, electron spin resonance, clay mineral diagenesis, fluid inclusions, pollen coloration, and vitrinite reflectance.

Pollen coloration and vitrinite reflectance are the most generally used. Pollen gradually changes color as it is heated, from colorless to yellow, orange, brown, and black. These can

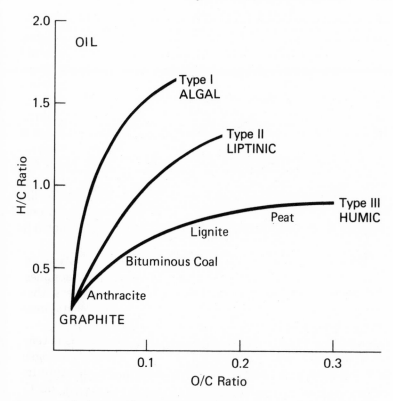

Figure 1 Chemistry of kerogens and coals.

be coded numerically and calibrated against temperature. Certain organic matter gradually becomes shinier as it is heated. The reflectivity of this vitrain can also be measured and calibrated. These techniques can be used on most source rocks and can thus give an estimate of their maturity (Fig.2).

Migration of Hydrocarbons

Geologists distinguish between two types of migration. *Primary migration* is the emigration of oil and gas from the

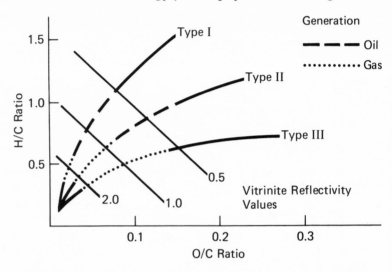

Figure 2 Relationship between hydrocarbon generation and vitrinite reflectivity.

source rock (clay) into permeable carrier beds. *Secondary migration* is the flow of hydrocarbons within permeable carrier and reservoir beds.

Secondary migration is well understood. Oil, gas, and water move through the porous rocks in response to buoyancy and to capillary and regional pressure gradients.

Primary migration is not well understood. Major questions still to be answered include: What was the chemical state of the emigrating fluids? Do mature oil and gas move from the source rock in discrete phases, or in solution in water, or do they emigrate in some transitional phase, i.e., *protopetroleum*?

What are the physical processes that cause hydrocarbons to emigrate from source beds? Straightforward compaction is inadequate as Figure 3 shows. Very little compactional porosity loss occurs at the depths and temperatures at which hydrocarbon generation takes place.

A possible mechanism is the expulsion of hydrocarbons as *protopetroleum*. Another mechanism may be the expulsion of hydrocarbons in solution, i.e., dissolved in water (derived from residual pore water, or from *structured* water in smectite clay

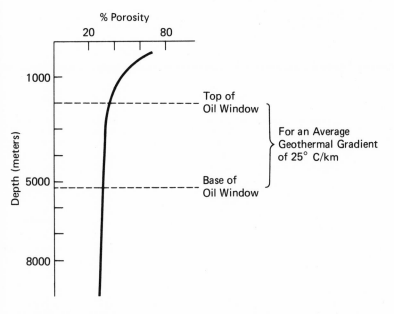

Figure 3 Clay compaction curve showing that there is little porosity loss through the oil window.

lattices); in solution of oil in gas; or attached to naturally occurring soaps (micelles). Two other mechanisms are: globules of oil in water, and continuous phase, i.e., three-dimensional continuity.

The actual mechanics of primary migration is largely of academic interest, and there are weighty arguments for and against the various processes outlined above. Current thought tends to favor the emigration of gas disolved in water. Solubilities of oil in water are negligible, however, even at high temperatures. Emigration of oil in a continuous phase is likely in rich source rocks, aided by structural water expelled from between the lattice layers of clay minerals as they recrystallize.

3

Geophysical Well Logs

Introduction

The study of rocks penetrated by a borehole is termed *formation evaluation*. This includes mudlogging, which is the examination of drilling mud circulated up from the bit. Gaseous and
liquid hydrocarbons can be detected and the lithology of the
formation identified from the rock cuttings. Formation evaluation also includes geophysical or wireline well logging. From
time to time as a well is drilled, various geophysical tools
(sondes) are lowered down the borehole. These record electric,
radioactive, acoustic, and other physical properties of the formation. These logs can be interpreted to determine lithology,
porosity, and the presence and amount of gas or oil.

Formation evaluation is a major branch of petroleum geology and engineering. In this book, emphasis is placed on the
geologic uses of well logs in lithology, identification, and facies
analysis. Porosity measurement and hydrocarbon evaluation
are only touched on briefly.

Geophysical Logging Introduction

Boreholes can produce a lot of subsurface information, particularly when extensive lengths of core are recovered. Because
coring is so expensive, it is more common, certainly in the oil

industry, to drill a hole with an ordinary bit and then to gain information about the penetrated formations by measuring their geophysical properties. This is generally done before setting casing, so several log-runs may be made during the drilling of a single well. The equipment that measures the geophysical properties of the rock is housed in a cylindrical sonde, which is lowered down the borehole on a cable. When the sonde reaches the bottom of the hole, the circuits are switched on and the various properties are measured as the sonde is drawn to the surface. The measurements are transmitted up the cable and are recorded on film or magnetic tape in the logging unit. In onshore operations the logging unit is generally mounted on a truck.

Many different parameters can be recorded, such as the resistivity, sonic velocity, density, and radioactivity of the rocks. These can be interpreted to determine the lithology of the penetrated formations, their porosity, and the nature of the fluid (oil, gas, or water) within the pores.

Zones of radioactive minerals can be located and many different types of radioactive minerals can be differentiated (such as thorium, uranium, and potassium). Coal seams can be located, and the various types of evaporite minerals can also be identified.

Some of the more common types of geophysical logging methods will now be described. Their uses are charted in table 6.

TABLE 6
THE USES OF THE MAJOR GEOPHYSICAL LOGS

	Lithology Determination	Porosity	Hydrocarbons
Electric Logs			
Spontaneous			
Potential (S.P.)	x	(perm)	
Resistivity			x
Radioactive			
Gamma	x		
Neutron		the	x } identify gas
Density		porosity x	
Sonic	x	logs x	
Caliper—measures hole size mechanically			
Dipmeter—a four arm microresistivity log			

The Spontaneous Potential Log

The Spontaneous Potential (S.P.) log is the oldest type of geophysical log. The first one was run over fifty years ago. The S.P. log records the charge set up between an electrode in a sonde drawn up the borehole and a fixed electrode at the earth's surface (measured in millivolts). It can be used only in open (i.e., uncased) holes filled with conductive mud, where there is a significant salinity difference between the mud and formation fluid.

The electric charge is basically related to the permeability of the formation. Deflection of the log to the left (negative on the conventional method of display) indicates porous sandstones and limestones; deflection to the right (positive on the conventional method of display) indicates impermeable shales or tight limestones and sandstones.

A sandline can be drawn through the lowest readings, and a shale base line through the highest. In this way, the vertical distribution of lithologies in a borehole can be found. The S.P. log can also be used to determine the resistivity of the formation water.

The Resistivity Log

The resistivity log records the resistance (in ohms) between positive and negative electrodes mounted on a sonde. The lateral distance from the borehole at which the resistivity of the formation is measured depends on the vertical interval between the two electrodes on the sonde. Widely spaced electrodes measure the resistivity far away from the borehole, where the formation is uncontaminated by drilling mud. Closely spaced electrodes measure resistivity close to the borehole where flushing has occurred.

Shales and salt-water-saturated porous formations have low resistivity. Tight formations and oil-saturated porous beds have high resistivities. Resistivity logs can locate hydrocarbon-bearing formations and can be used to find coal beds, which are also highly resistive.

The Gamma Log

The gamma log, one of the most common of the three major types of radioactivity logs, records the radioactivity of the formation. It can be used in both cased and uncased true-gauge holes.

The major radioactive element in sediments is potassium, which is found largely in clay minerals. An increase in the gamma reading in the formation generally indicates shale; deflection to the left (low radioactivity) indicates sand or carbonate. Sand and shale base lines can be constructed in the same way as for the S.P. log. The gamma log is more useful than the S.P. log because the gamma log can differentiate shale from limestone and sandstone irrespective of their permeability.

Fundamentally, this is a lithology and correlation log. More specialized gamma logs can be used to differentiate radioactive elements such as thorium.

The Neutron Log

With the neutron log, the sonde bombards the formation with neutrons. The gamma rays that are emitted are recorded on the sonde. Radiation intensity is essentially proportional to the amount of hydrogen in the formation. This is related largely to the amount of fluid (water, oil, or hydrocarbon gas). Hence, the neutron log basically records porosity. It gives too low a reading in gas reservoirs.

The Density Log

With the density log the sonde emits gamma radiation. The gamma rays are scattered on reaching electrons in the formation, and the returning gamma rays are recorded on the sonde.

The returned radiation is proportional to the number of electrons in the formation, which is proportional to the bulk density of the rock. Porosity may be calculated from this formula:

$$\phi = \frac{\rho_{ma} - \rho_b}{\rho_{ma} - \rho_f}. \tag{2}$$

where: ϕ represents porosity, ρ_b is the density recorded on the log, ρ_{ma} is the density of the matrix (approximately 2.65 for sandstones and 2.71 for limestones), and ρ_f is the density of the pore fluid (approximately unity).

The log records density in grams per cubic centimeter (gms/cc). The density and neutron logs are important porosity-measuring logs and are very useful when plotted together. This is because in the presence of hydrocarbon gas, the neutron log reads too low a porosity while the density log records too high a value. Lateral separation of neutron and density log curves can thus be used to detect gas zones in the borehole. (see Fig. 4)

The Sonic Log

The sonic log measures the interval transit time (Δt) for a compressional sound wave travelling through the formation. This is measured in microseconds per foot. It works only in true gauge open hole. The sonic log responds to variations in lithology and to porosity.

The basic formula is:

$$0 = \frac{\Delta t \log - \Delta t_{ma}}{\Delta t_f - \Delta t_{ma}}. \tag{3}$$

where: Δt log is measured from the log, Δt_{ma} is the transit time for the matrix material, and Δt_f is the transit time for the pore fluid (about 189 sec/foot). Δt_{ma} varies from 55–51 sec/foot for sandstones and 47–43 sec/foot for limestones.

The sonic log is used, therefore, both to identify porosity and lithology. It is particularly good for identifying coal beds

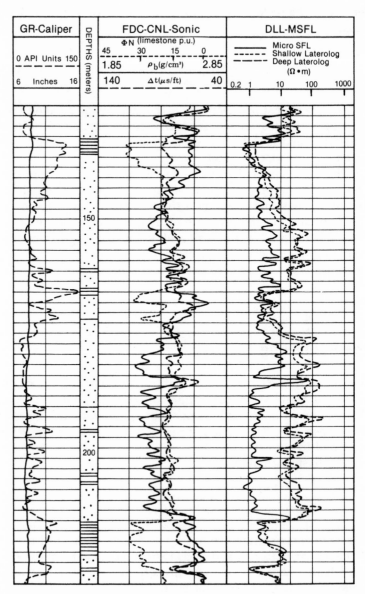

Figure 4 Suite of logs through shales and gas-bearing sands from the Triassic of Algeria. (Courtesy of Schlumberger)

because of their very high sonic transit times. It is less accurate at measuring porosities than the neutron and density logs, because of the variation of sonic transit time with lithology.

Geophysicists use the sonic log to calibrate seismic time sections against depth, and hence determine the acoustic velocities of the various formations. Figure 4 illustrates a suite of logs run through an interval of shales and gas-bearing sands.

The Dipmeter Log

The dipmeter log measures the dip of strata adjacent to the borehole. Basically, it consists of four microresistivity logs recorded at right angles to one another. Correlations on the curves are picked by computer, and dips are calculated from them.

The correlation interval within which the computer looks for correlation can be specified, as can the step distance that the computer examines for each correlation. Generally, correlation interval and step distance should be equal. For structural dip determination, 12 × 12 foot or 9 × 9 foot plots are used. For sedimentary dips a 3 × 3 foot may suffice, but ideally 1 × 1 foot or 0.5 × 0.5 foot are necessary. Search angle may also be specified, and the computer will remove structural tilt. Search angle should be less than 35° when looking for sedimentary dips.

Dipmeter data are generally presented in a *tadpole* plot. Four basic tadpole plot motifs can be recognized in Figure 5: upward-increasing (blue), upward-decreasing (red), steady (green), and random (bag o'nails). They may be due to all sorts of geological phenomena.

The dipmeter should never be interpreted in isolation but only with all other available data. When used carefully, it can identify folds, faults, and unconformities, as well as smaller scale features such as cross-bedding. Knowledge of cross-bed-

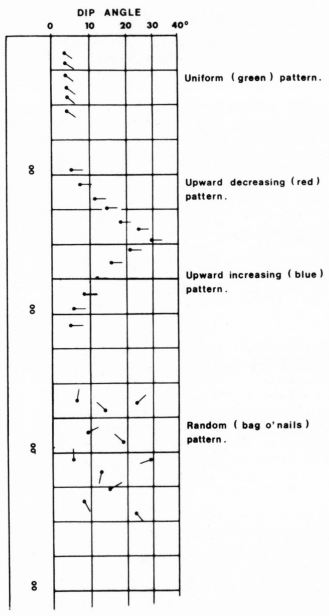

Figure 5 Basic dipmeter motifs. (From Richard C. Selley: Ancient Sedimentary Environments, Second Edition. ©1970, 1978 by Richard C. Selley. Used by permission of the publisher, Cornell University Press; and used by permission of the originating British publisher, Chapman & Hall Limited)

ding directions in a sand body can be used to predict its orientation. This is very useful when an aquifer or hydrocarbon reservoir is found in channel or sandbars.

Geophysical Logging Summary

The logging and interpretation of boreholes is a far more complex discipline than the preceding pages can indicate. It is a study that necessitates a knowledge of geology, physics, electronics, and computer technology. At their simplest well logs provide the basis for the mapping of the aerial and vertical distribution of strata and their contained fluids. At their most complex geophysical logs can be used to measure the different proportions of the various minerals (such as biotite and muscovite mica) that may be present in rocks.

Composite well logs, which combine geophysical logs with geologic and palaeontologic studies of cores and cuttings, are the basis for the subsurface mapping of structure, sedimentary facies, and palaeogeography.

Integrated Facies Analysis

When integrated, the dipmeter, gamma, and S.P. logs can be of great use in subsurface facies analysis, because the S.P. and gamma logs are essentially vertical profiles of grainsize in clastic sequences. Grainsize profiles are environmentally diagnostic, though no profile is unique to any environment.

When coupled with the presence or absence of glauconite and carbonaceous detritus in well samples, however, grainsize profiles can be very useful. Glauconite indicates a marine environment, and the presence of carbonaceous detritus indicates a poorly winnowed environment, marine or continental. Hence, when glauconite is present and carbonaceous detritus is absent, the environment is marine, and well winnowed. This is typical of beach, barrier bar, and marine shoal sand. If glauconite and carbonaceous detritus are both present, the environment is marine, and poorly winnowed, i.e., deep sea and turbidite. If carbonaceous detritus is present and glauconite is absent, the environment is poorly winnowed, possibly nonmarine. This is typical of deltas, lakes, and alluvial flood plains. When carbonaceous detritus and glauconite are both absent, the result is nondiagnostic.

Coupled with S.P. or gamma profiles, this approach may identify many environments (see Fig. 6). Figures 7 and 8 illustrate typical log responses for deltaic and deep sea sands.

Core and paleontological data should always be studied before this method is used.

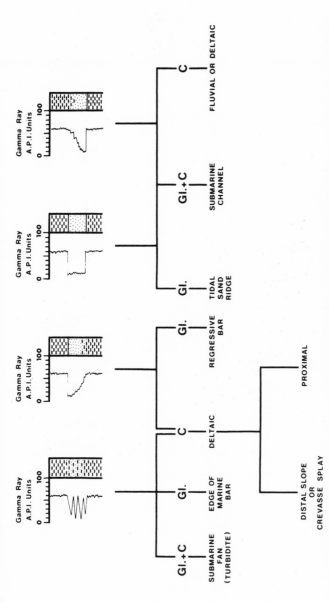

Figure 6 Characteristic log motifs. From left to right: thinly interbedded sand and shale; an upward coarsening profile with an abrupt upper sand: shale contact; a uniform sand with abrupt upper and lower contacts, and furthest right, an upward fining sand: shale sequence with an abrupt base. None of these log patterns are diagnostic on their own. Coupled with data on the distribution of glauconite and carbonaceous detritus however, they define the origin of a number of sand bodies. (From Selley 1976, courtesy AAPG)

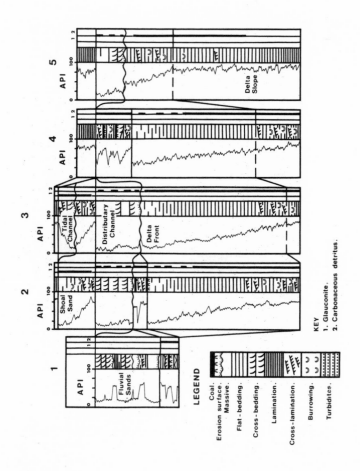

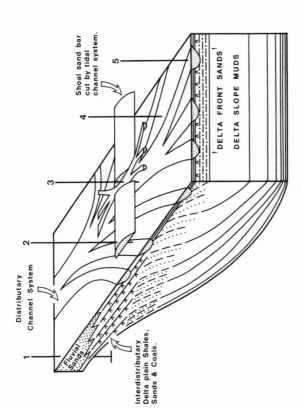

Figure 7 Cartoon to indicate the sand bodies and log characteristics of deltaic sedimentary facies. When core material is available, suites of sedimentary structures are often diagnostic. Even without core material, however, depositional environment may be detected from log motif (grainsize profile) and the distribution of glauconite and carbonaceous detritus. (From Richard C. Selley: Ancient Sedimentary Environments, Second Edition. ©1970, 1978 by Richard C. Selley. Used by permission of the publisher, Cornell University Press; and used by permission of the originating British publisher, Chapman & Hall Limited)

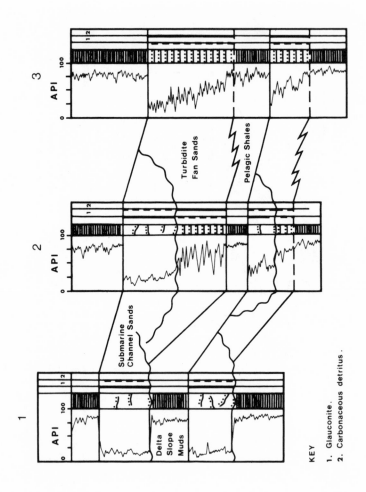

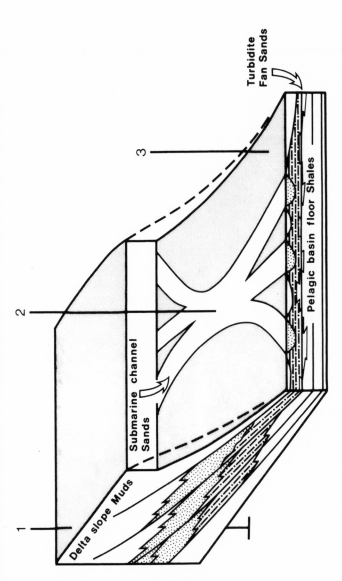

Figure 8 Cartoon to illustrate the sand bodies and log characteristics of deep sands. These may occur at the feet of deltas and the base of submarine fault scarps. (From Richard C. Selley: Ancient Sedimentary Environments, Second Edition. ©1970, 1978 by Richard C. Selley. Used by permission of the publisher, Cornell University Press; and used by permission of the originating British publisher, Chapman & Hall Limited)

4

Reservoir Rocks I

Introduction: Porosity and Permeability

A reservoir rock must have two properties: *porosity* (there must be void spaces within the rock to store hydrocarbons) and *permeability* (fluid must be able to flow through the rock). It is essential for the pore spaces in a reservoir rock to be connected. It is important to separate total porosity from effective porosity. Effective porosity refers to the amount of total pore volume that is actually interconnected and thus stores and gives up fluid. Porosity is conventionally expressed as a percentage:

$$\text{porosity } (\%) = \frac{\text{volume of pore space}}{\text{volume of rock sample}} \times 100. \qquad (4)$$

Most reservoir rocks have porosities of 15–30%. Porosities of less than 10% are generally subeconomic unless there are other favorable parameters (e.g., vast reservoirs, favorable geographic location, easy tax terms, etc.).

Porosity is measured in one of three ways. One method is the direct measurement of a core, or core plug, in a laboratory; another method is the use of geophysical logs in boreholes, sonic velocity, neutron, or bulk density. Porosity can also be measured from seismic interval velocities. This last method can be difficult unless the lithology is already known from drilling.

Porosity Classification

Porosity is classified as primary and secondary according to its time of formation, as is shown in table 7.

TABLE 7
A CLASSIFICATION OF POROSITY

Type		Origin
Primary or depositional	Intergranular or interparticle Intragranular or interparticle	Sedimentation
Secondary or post-depositional	Intercrystalline Fenestral	Cementation
	Vuggy Moldic	Solution
	Fracture	Tectonics Diagenesis Dehydration Compaction

Sandstones generally have primary intergranular porosity, while limestones generally have secondary porosity. Igneous and metamorphic rocks (basement) are generally tight, though they may have fracture or solution porosity beneath unconformities caused by weathering.

Fracture porosity may occur in any brittle rock—igneous, metamorphic, sandstone, limestone, or shale. Certain common petrophysical properties for various rocks are shown in Figure 9.

Permeability

Permeability is conventionally expressed by Darcy's Law. It is measured from cores by recording the rate of flow and pressure drop when a fluid of known viscosity flows through a rock sample of known length and cross-sectional area (Fig. 10). Then:

$$Q = \frac{KDA}{\mu L}.\tag{5}$$

where: Q is the rate of flow in cc/sec, K is the permeability, D is $P_1 - P_2$, and μ is the viscocity of the fluid.

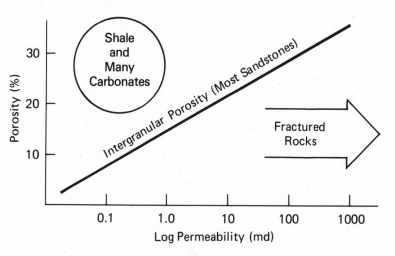

Figure 9 Petrophysical properties of some rocks.

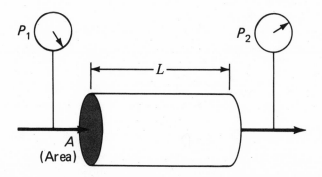

Figure 10 Diagram depicting that permeability is measured by recording the rate of flow and pressure drop.

The unit of permeability is the Darcy. One Darcy is the value that allows a fluid of one centipoise viscosity to flow at one cm/sec for a pressure gradient of one atmosphere per cm. Most rocks have permeabilities of considerably less than one Darcy, so the millidarcy (md) is commonly used.

Most reservoir rocks have permeabilities in the range of 100–500 md, though higher values may occur. As with porosity, lower permeabilities may be economic, especially for gas production because of its low viscosity.

Sandstone Reservoirs

Nearly half the world's oil and gas is produced from sandstone reservoir rocks. Sandstone is defined as a sedimentary rock formed of sand-sized grains (2.0–0.0625 mm diameter). Leaving aside carbonate sands, most sandstones are referred to as terrigenous (land derived) or siliciclastic (composed largely of silica minerals). Terrigenous sands are produced by the weathering and erosion of igneous, metamorphic, or pre-existing sedimentary rocks. The main mineral components are quartz (SiO_2), feldspar (calcium-, sodium-, and potassium-alluminosilicates), mica (potassium-, magnisium-, and iron-hydrated alluminosilicates), mafic minerals (olivine, horneblende and augite—magnisium and iron silicates), and heavy minerals (iron ores, zircon, etc.).

As a sediment is formed by weathering, erosion, transportation, and deposition, it gradually becomes more mature. One type of maturity is chemical maturity where unstable minerals (mafics, feldspar, etc.) dissolve out resulting in a relative increase in the proportion of stable mineral grains. A second type of maturity is physical maturity where the sand becomes better sorted, clay is winnowed out, and grains become better rounded. As physical maturity increases, porosity and permeability increase.

Sandstones are generally classified into four main groups according to their physical and chemical maturity (Fig. 11). Each sandstone type has a different potential for being a reservoir. As a sand is buried it gradually loses porosity by compaction due to minor adjustments of grain packing and by cementation due to the precipitation of minerals within pores. The three main chemical cements are: quartz (SiO_2), calcite ($CaCO_3$), and clay minerals. Both quartz and calcite are derived from pore fluids expelled from adjacent compacting clays. Clay minerals are derived from the solution and reprecipitation of chemically unstable mineral grains in the sand. Figure 12 shows what happens when the four major sandstone types are buried.

Depositional Processes and Environments

Basically sand is deposited by two different physical processes: *traction currents* and *turbidity currents*.

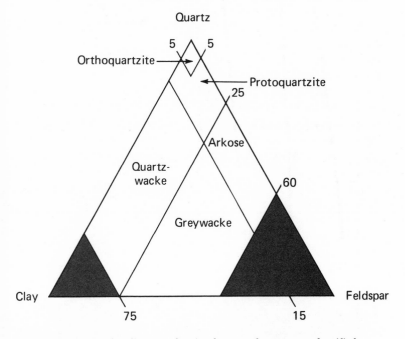

Figure 11 Triangular diagram showing how sandstones are classified according to their percentage composition of quartz and feldspar grains and clay matrix.

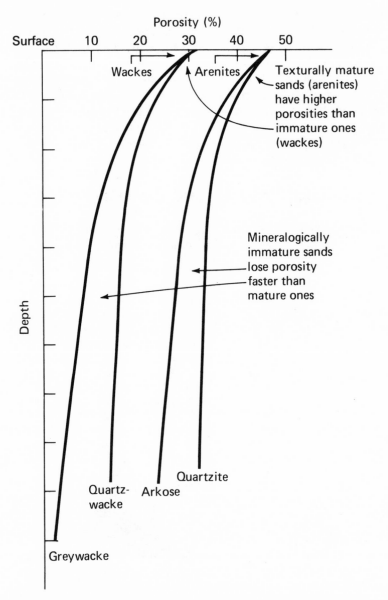

Figure 12 Diagram to show how sandstone types lose porosities at different rates during burial.

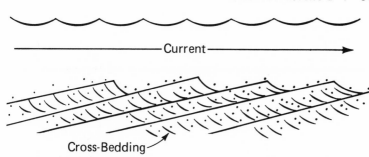

Figure 13 Diagram depicting megaripples depositing cross-bedding.

Traction Currents

Traction currents are horizontally flowing water (or air) that cause sand particles to bounce along in a down-current direction. At low velocities, migrating ripples deposit cross-lamination; at higher velocities, megaripples (up to several meters in height) deposit cross-bedding (Fig. 13). Traction currents occur in channels, in sand deserts, and on marine shelves and coasts.

Turbidity Currents

Turbidity currents are caused by a dense sediment-laden mass of water flowing down a slope beneath less dense clear water. At the foot of the slope the sediment settles out of suspension to form an upward-finding graded bed, termed a turbidite. Turbidites are characteristic of deep-sea sands (Fig. 14).

Sands may be deposited in a whole range of sedimentary environments (Fig. 15), which deposit characteristic facies. Facies is a mass of sediment with characteristic geometry, lithology, sedimentary structures, fossils, etc. Subsurface facies analysis attempts to deduce the depositional environment of a reservoir rock using cores, geophysical logs, palaeontology, and seismic stratigraphy. It enables predictions to be made of the geometry and orientation of porous permeable beds. Table 8 illustrates how sedimentary environments are commonly classified.

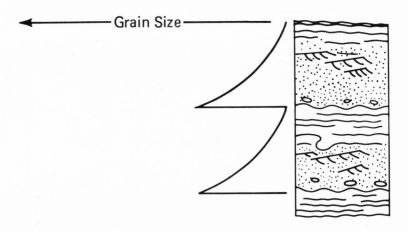

Figure 14 Diagram depicting coarser particles settling out of suspension first. The finer particles are deposited on top.

TABLE 8
SEDIMENTARY ENVIRONMENTS

Continental	⎰ fluvial ⎱ eolian lacustrine
Shoreline	⎰ deltaic ⎱ barrier island and lagoon

Marine	⎰ shelf
	  reef
	  deep-sea sand (submarine channel and turbidite fan)
	⎱ pelagic (basin floor mud)

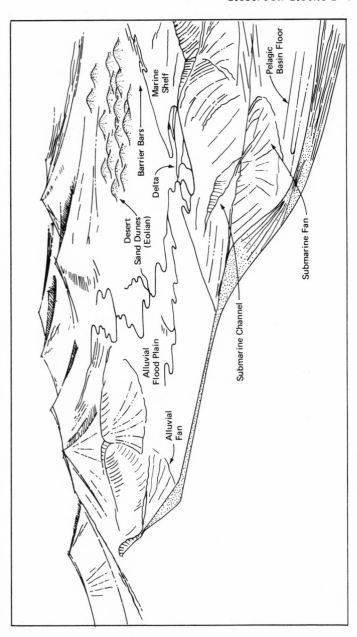

Figure 15 Geophantasmogram to illustrate the various environments in which sedimentation takes place.

5

Reservoir Rocks II: Carbonates

Introduction

Carbonate rocks (limestones and dolomites) differ from terrigenous rocks in a number of ways. Carbonate sediment is intrabasinal in origin. There are a number of grain types that form in specific environments. Carbonate sediments have higher porosities and permeabilities than terrigenous sands, especially in reefs (see Fig. 16). Carbonate minerals are less stable than quartz.

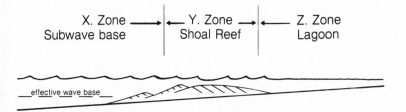

Figure 16   Carbonate shelf facies model. (From Irwin 1965, courtesy AAPG)

Because of these differences, the techniques of subsurface facies analysis and of porosity location are very different. For sandstones, subsurface facies analysis is based on studying grainsize profiles and sedimentary structures. Lithology is of little diagnostic value. In the quest for porosity, the effect of facies is generally greater than diagenesis. For carbonates, subsurface facies analysis is based on petrography. In the quest for porosity the effects of diagenesis is generally more important than facies controlled primary porosity.

Carbonate Mineralogy and Components

Most modern carbonate sediments are composed of aragonite (the orthorhombic variety of $CaCO_3$). Most ancient limestones are made of calcite (the hexagonal variety of $CaCO_3$). Early diagenetic (and hence porosity) changes in carbonates are due to the reversion of aragonite to calcite.

Dolomite $[CaMg(CO_3)_2]$ occurs in two ways: penecontemporaneous dolomite tends to be microcrystalline and has a low permeability unless fractured. Secondary replacement dolomite tends to be macrocrystalline and, because there is a bulk volume reduction when calcite is replaced by dolomite, often has secondary intercrystalline porosity.

The various types of carbonate grain are shown in table 9.

TABLE 9
CLASSIFICATION OF THE COMPONENTS OF CARBONATE ROCKS

Grains	detrital grains
	  lithoclasts
	  intraclasts
	skeletal grains
	peloids (including faecal pellets)
	lumps
	  composite grains (grapestone)
	  algal lumps

Matrix	micrite clay
Cement	sparite
Pores	voids

Most skeletal and oolitic limestones were deposited in high energy shoal and bar environments: Faecal pellet deposits characterize lagoons and tidal flats. Lime mudstones (micrites) tend to be basinal.

Some possible origins of micrite are (1.) the disintegration of calcareous algae into individual crystallites, (2.) the organic destruction of calcareous skeletons, ranging from coral-crunching parrot fish to boring algae, (3.) the inorganic destruction of skeletons by wave action, and (4.) the direct precipitation (whitings).

Carbonate Environments and Facies

Irwin (1965) proposed a basic carbonate facies model based on the interplay of the sea floor with sea level and effective wave base (Fig. 16). The close control between environment, facies, and grain type is summarized in table 10. Examples of typical X-, Y-, and Z-zone lithologies are given in Figures 17, 18, and 19.

TABLE 10
RELATIONSHIPS BETWEEN ENVIRONMENT FACIES
AND GRAIN TYPES

Environment	Facies	Grain Types
X-zone (deep basinal, subwave base)	Lime mudstones and chalks	Micrite, with minor skeletal grains
Transition	Wackestone - packstone	
Y-zone, (shallow high energy shoal and reef)	Grainstone, boundstone, packstone	Skeletal grains, ooliths, minor intraclasts, and composite grains
Transition	Wackestone	
Z-zone (shallow low energy lagoon, tidal flat and sabkha	Pellet mudstones, stromatolites, primary dolomite, and anhydrite	Faecal pellets, skeletal grains, and micrite

Figure 17 Basinal lime mudstone (Cretaceous, Abu Dhabi). This is an example of a typical X-zone deposit.

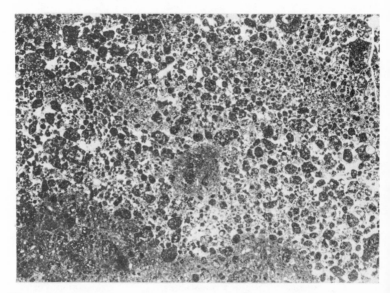

Figure 18 Skeletal lime sand [with minor amounts of pellets (Miocene, Libya)]. A typical example of a high-energy Y-zone deposit.

Figure 19 Faecal pellet lime sand. Upper Jurassic, Southern England. A typical example of a Z-zone lagoonal or intertidal flat deposit.

Reefs

Strictly speaking a reef is something on which a ship may be wrecked. Usually, in geology, the term reef is applied to a lens of limestone formed by in situ colonial organisms. A bioherm is a mound of organic origin, and a biostrome is a sheet of organic detritus. The term build-up is a useful non-genetic term.

Modern Reefs

A modern reef is a rigid framework of corals and coralline bryozoa encrusted and bound together by algae. This is a major source of coralgal sand, as well as of in situ limestone. Modern reefs are found in shallow (50 m), warm (20 °C), clear, salty [27,000–40,000 parts per million (ppm)] water. Their geometry may be linear, as in fringing reefs, found against arid coasts, or barrier reefs, adjacent to humid coasts. Pinnacles or patch reefs are scattered randomly in lagoons or on open shelves. Atolls are found at the edge of Pleistocene bevelled carbonate platforms or on drowned volcanoes as part of the Darwinian sequence.

Ancient Reefs

Ancient reefs are composed not only of corals, bryozoa, and algae, but also of stromatoporoids and certain groups of brachiopods and lamellibranchs (e.g., the Rudists). When they were formed, their environment was apparently the same as today. Their geometry may be linear as in fringing reefs, but these are rare (low preservation potential?), or shelf margin barrier reefs, which are common, e.g., Guadalupe, Texas and Canning basin, Australia. Pinnacles are also well known, e.g., Devonian of Alberta. Rare examples of atolls are the Permian Horseshoe Atoll of Texas, and the Cretaceous El Abra reef complex of Mexico.

Economic Significance of Reefs

Some reefs are susceptible to replacement by telethermal lead-zinc sulphides. Examples include the Carboniferous of Ireland and England; Slave Creek, Canada; and Red Sea Miocene. Reefs may be reservoirs for oil and gas especially when associated with black shales as source rocks and evaporites as cap rocks. Examples are: Devonian of Alberta, Permian of Texas, Upper Cretaceous and Palaeocene of Libya (Figs. 20 and 21).

Subsurface Facies Analysis

The geometry of a reef is linear or round and may be identified from gravity or seismic surveys. The lithology, biolithite or boundstone, is not always easy to identify in cuttings or cores, especially if recrystallized or replaced by dolomite. Major talus bedding may be detectable on the fore-reef from seismic or dipmeter logs and the drape of onlapping shales may also be detectable. Dipmeter logs may be used to define reef geometry (Fig. 22).

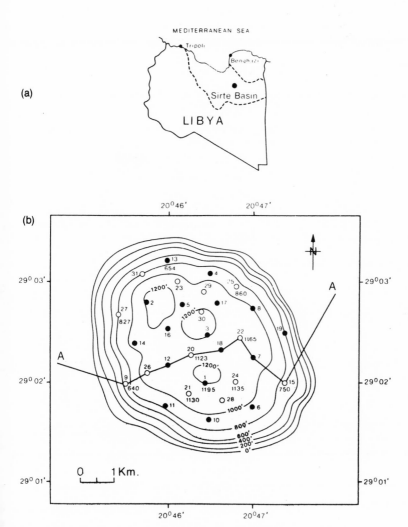

Figure 20 (a) Map of Libya showing location of Idris (Intisar) fields.
(b) Isopach map of the 'A' reef. (After Terry and Williams 1969; courtesy
Institute of Petroleum, London)

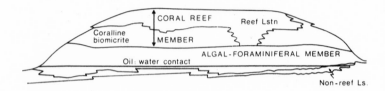

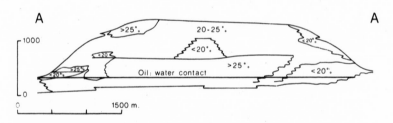

Figure 21 Cross sections through the Idris 'A' bioherm. Note how facies are unrelated to porosity distribution. This is common in many carbonate reservoirs. (After Terry and Williams 1969; courtesy Institute of Petroleum, London)

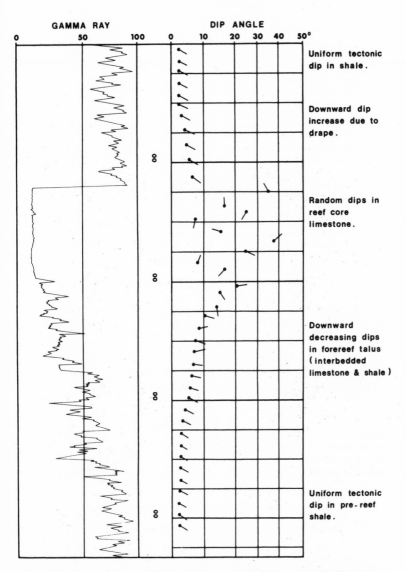

Figure 22 Gamma log and dip motif for a well drilled on the SE flank of a reef. (From Richard C. Selley: Ancient Sedimentary Environments, Second Edition. ©1970, 1978 by Richard C. Selley. Used by permission of the publisher, Cornell University Press; and used by permission of the originating British publisher, Chapman & Hall Limited)

6

Evaporites

Introduction

Another major group of sedimentary rocks are the evaporites. Unlike the sandstones and limestones, they do not form reservoirs. They are very important to petroleum exploration, however, because evaporites make excellent cap rocks, generate traps, may act as source rocks, and confuse geophysicists.

The evaporites are a large group of crystalline (as opposed to granular) sedimentary rocks, which were once thought to have occurred by the evaporation of sea water. Many geologists now believe, however, that these rocks are formed, in part at least, by the replacement of previously formed carbonate sediment (see table 11).

TABLE 11
COMPOSITION AND MINERALOGY OF EVAPORITES

Mineral	Formula	Density g/cc
Anhydrite	$CaSO_4$	2.960
Gypsum	$CaSO_4 \cdot 2H_2O$	2.320
Halite (rocksalt)	$NaCl$	2.165
Sylvite	$K\ Cl$	1.984

An evaporite rock, with a mixture of the minerals listed in table 11, will have an average density of 2.1–2.6, which is less than common values for sandstones (2.65), limestones (2.71), and dolomites (2.87). This explains why evaporites deform and move up through overlying denser sediment and why they have significantly different acoustic velocities.

Occurrence and Formation of Evaporites

Evaporites occur associated with shallow marine limestones, dolomites, continental red shales, and sands (e.g., the Permian of the North Sea). Sedimentation is generally cyclic, reflecting gradual increases in salinity followed by abrupt returns to normal sea water conditions (Fig. 23). Because of this mode of oc-

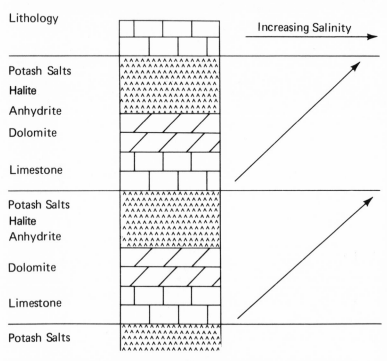

Figure 23 Section depicts cyclical increases in salinity of evaporites.

currence, many geologists believed that evaporites formed as a result of the progressive drying up of barred coastal basins, followed by a renewed influx of normal sea water. This is the so-called *evaporating dish* mechanism (Fig. 24).

This mechanism explains both the cyclic stratigraphy of evaporite sequences and the fact that, within each cycle, salinity increases towards the basin center. Objections to this mechanism are the improbability of great depths of sea water evaporating to deposit one cycle, and the fact that, proportionally, the various salts found in an evaporite cycle are different from those present in sea water.

Recent research into sedimentation of modern, arid salt-marsh (Arabic: sabkha) coasts has shown evaporites forming in a very different way. Here carbonate muds are deposited in hypersaline lagoons and on tidal flats. Evaporation due to the hot sun draws lagoonal brines along and up into the intertidal and supratidal lime muds. As the brines evaporate, dolomite, anhydrite, and other minerals form displacive nodules, and replace the original carbonate grains (Fig. 25).

Many geologists now believe that the vast thicknesses of evaporites distributed around the world were not formed on the beds of deep hypersaline seas, but by the to and fro migration of sabkha shorelines across a shallow, gradually subsiding shelf.

Phase I: high water level; normal salinity in lagoon

Sea
Level

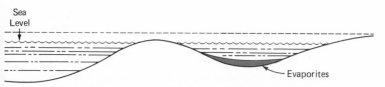

Evaporites

Phase II: sea level drops; lagoon is cut off from sea so salts evaporate out

Figure 24 The classic *evaporating dish* mechanism proposed for the origin of evaporite cycles.

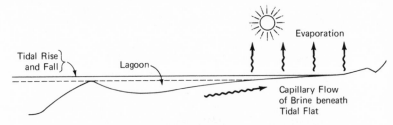

Figure 25 Cross section through a lagoon and salt marsh (sabkha) in a desert climate showing how brines, concentrated in the lagoon, are drawn beneath the tidal flat where they evaporate and precipitate their salts, often replacing carbonate muds.

Significance of Evaporites to Petroleum Exploration

Evaporites as Cap Rocks

The essential property of the cap rock located above a hydrocarbon reservoir is that it must be impermeable. Evaporites make the perfect seal, not only because they are impermeable per se, but because, unlike lithified shales, they deform plastically, not by fracturing.

Evaporites as Trapping Mechanisms

Because they are less dense than most sediments, evaporites tend to move upwards through younger cover, forming linear salt walls (or pillows) and domes (or diapirs). Movement may be slow, shown by syndepositional thinning of overlying strata, or rapid, accompanied by piercing of strata. Movement is often accompanied by a radial fault pattern and extensive fracturing of brittle rocks (e.g., the Ekofisk fields of the North Sea).

Salt doming (halokinesis) can cause all sorts of potential hydrocarbon traps (Fig. 26). Examples of major salt dome pro-

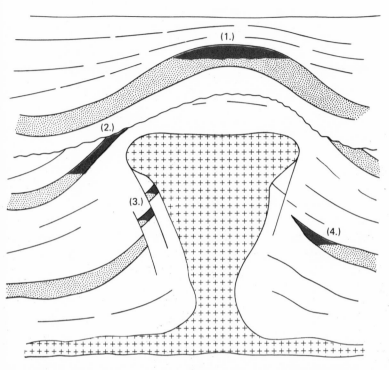

Figure 26 Cross section through a salt dome to show some of the different types of traps that may occur: (1.) crestal dome; (2.) truncation; (3.) fault trap; (4.) sedimentary pinch out.

vinces include: the U.S. Gulf Coast, where the Louann salt (Jurassic) has moved up through a Tertiary clastic wedge; the Persian Gulf, where the Hormuz salt (Cambrian) has pierced Mesozoic and Tertiary carbonates; and the Ekofisk play of the North Sea, where Permian Zechstein evaporites have domed and fractured Cretaceous chalk.

Another type of salt feature occurs where, in an area of structural movement, layers of evaporites have acted as glide surfaces allowing overlying sediments to move disharmonically with respect to subsalt rocks. The horizontal salt glide surfaces are known as zones of decollement. Classic examples occur in Iran (Fig. 27).

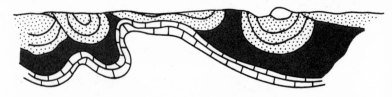

Figure 27 Cross section to show the disharmonic nature of folding in strata above and below evaporite beds (shown in black.) Based on Masjid-i-Sulaiman oil field, Iran. (After G.M. Lees, 1953)

Evaporites as Source Rocks

The conditions that favor evaporite formation also favor the preservation of organic matter. Thus, many evaporites are interlaminated with kerogen, principally of algal origin. It has been suggested that these may act as hydrocarbon source rocks, though the mechanics of primary migration are obscure and probably more complex than for normal clay source beds.

7

Hydrocarbon Traps

Introduction

Source rock, reservoir, and seal must be arranged in such a geometry that hydrocarbons are trapped, to prevent their natural tendency to move up and be dissipated at the surface of the earth.

Basic Parameters of a Trap

Figure 28 illustrates the basic morphology and terminology of an antclinal trap. Note also that the gross pay is the vertical thickness from the top of the reservoir to the oil-water contact. The net pay, or net effective pay, is the aggregate producible thickness of the reservoir within the gross pay.

Classification of Hydrocarbon Traps

Structural traps are due to postdepositional tectonic processes (e.g., folds and faults). *Stratigraphic traps* are due to syndepositional sedimentary processes (e.g., reefs) or postdepositional diagenetic ones (e.g., dolomitization). *Hydrodynamic traps* are caused by flowing water. Generally they only occur as an element of a combination trap. *Combination traps* are caused by a combination or combinations of structural, stratigraphic, and hydrodynamic processes.

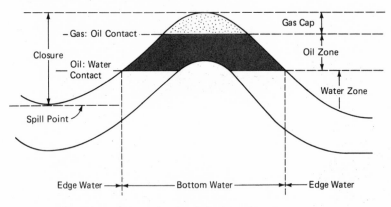

Figure 28 Nomenclature of a simple hydrocarbon trap.

Structural Traps

There are three main types of structural traps. One type is the anticline, or upfold of rock, which is divisible into compressional (due to crustal shortening) and compactional (due to compaction over basement highs of palaeotopographic or tensional fault block origin). These are illustrated in Figure 29.

A second type of structural trap is the fault trap, where porous, permeable reservoir beds are faulted against impermeable beds. Some fault planes seal, others act as permeable conduits. Faults are one of three main types: normal, which are due to tension; reversed (or thrust), which are due to compression; and tear (or wrench), which are due to horizontal movement (see Fig. 30).

Normal and reversed faults may generate traps, but tear faults seldom do, though wrenching may generate anticlines parallel or oblique to major shear zones (Fig. 31). Faults may be due to regional crustal compression and tension, or due to local forces, such as salt movement and clay compaction.

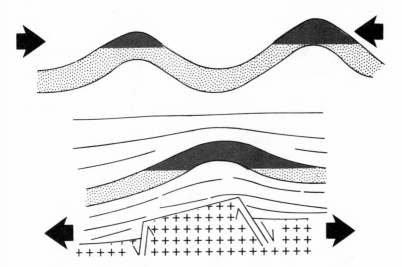

Figure 29 Diagrams depict compressional and compactional anticlines.

The third type is the growth fault, which shows sediment thickening on the downthrown side. This demonstrates synchronous movement. They are commonly produced either by compacting delta muds, or by prolonged movement of basement blocks. Traps may occur both against the fault and in adjacent *rollover* anticlines (e.g., the Niger delta) (Fig. 32).

Anticlinal, fault and other types of traps are associated with salt domes as shown in chapter 6.

Hydrodynamic Traps

The downward flow of water in a permeable bed may cause oil to be trapped in flexures which lack vertical closure. This type of pure hydrodynamic trap, illustrated in Figure 33, is virtually unknown, but many anticlinal oil fields, especially those above sea level, show tilted oil-water contacts due to flow of bottom water (Fig. 34). Tilted oil-water contacts may also be due to facies changes, or the tilting of fields in which the water zone has become cemented and tight.

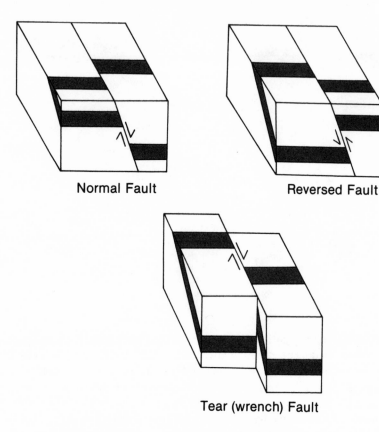

Normal Fault   Reversed Fault

Tear (wrench) Fault

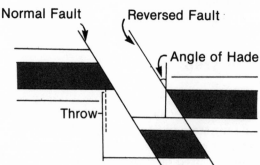

Figure 30 Diagrams depict the three main kinds of faults and fault terminology.

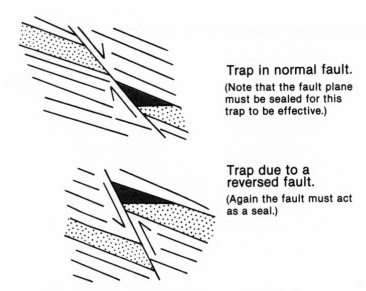

Trap in normal fault.
(Note that the fault plane must be sealed for this trap to be effective.)

Trap due to a reversed fault.
(Again the fault must act as a seal.)

Figure 31 Diagram depicts trap in normal and reversed faults.

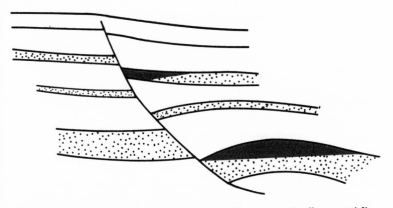

Figure 32 Diagram depicts a growth fault with associated rollover anticline. Oil is trapped both against the fault and within the anticline.

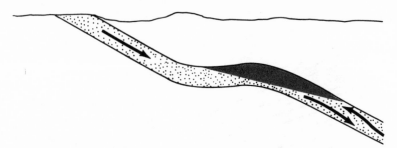

Figure 33 Cross section to illustrate oil trapped hydrodynamically by flowing water. Pure hydrodynamic traps like this are very rare.

Figure 34 Diagram depicting tilted oil-water contacts due to hydrodynamic flow.

8

Stratigraphic Traps

Introduction

A stratigraphic trap may be defined as one in which the chief trap-making element is some variation in the stratigraphy or lithology (or both) of the reservoir rock, such as a facies change, variable local porosity and permeability, or an up-structure termination of the reservoir rock, irrespective of the cause (Levorsen, 1967).

Data from Halbouty et al. (1970) show that only 13% of giant oil and gas fields are stratigraphic traps and that a further 11% are combination traps. This probably reflects our inability to locate stratigraphic traps, rather than the relative abundance of different trap types (see table 12).

Stratigraphic Trap Exploration Techniques

Traditionally, stratigraphic traps were found relatively late in the exploration of a basin when subsurface data were available from wells drilled to test for structural traps. Methods of stratigraphic trap location are either direct or indirect.

Direct Methods of Stratigraphic Trap Location. Direct methods of stratigraphic trap location are principally geophysical. These include gravity and seismic prospecting for reefs,

TABLE 12
A CLASSIFICATION OF STRATIGRAPHIC TRAPS

Traps unassociated with unconformities	$\left\{\begin{array}{l}\text{depositional traps} \\ \quad\text{channels} \\ \quad\text{sand bars} \\ \quad\text{reefs} \\ \text{diagenetic traps—poro/perm change}\end{array}\right.$
Traps associated with unconformities	$\left\{\begin{array}{l}\text{supraunconformity—palaeogeomorphic trap} \\ \text{subunconformity—truncation trap}\end{array}\right.$

Source: From Rittenhouse (1972)

and seismic prospecting for unconformity and sand body traps. With increased seismic resolution, this technique is of ever-increasing importance.

Indirect Methods of Stratigraphic Trap Location. Indirect methods of stratigraphic trap location consist mainly of subsurface facies analysis, which integrates log character analysis, dipmeter interpretation, petrographic core studies, palaeontology, and regional palaeogeography.

The various types of stratigraphic traps will now be outlined.

Unconformity-Associated Traps

Stratigraphic traps occur both above and below unconformities. In either case, they are generally overlain by transgressive marine shales, which act as the seal and often also as the source bed.

Supraunconformity Traps

Supraunconformity traps are generally sand reservoirs. Three main types may be defined as regional pinchouts, palaeochannels, and strike valley sands (Fig. 35). Palaeochannels and strike valley sands are generally referred to as palaeogeomorphic traps. Notable examples of supraunconformity traps include the strike valley sands of Oklahoma and New Mexico and channel sands, such as those of the sub-Cretaceous unconformity in many of the Rocky Mountain basins.

Subunconformity Traps

Subunconformity traps may occur in a diversity of rock types including sandstone, limestone, and various basement rocks (Fig. 35). In many such traps, secondary porosity, due to fracturing and solution (epidiagenesis), plays a major role in reservoir development. Examples of these traps include the Mississippi Carbonate fields trapped beneath the Cretaceous unconformity of Saskatchewan; the Brent fields of the North Sea, actually a combination trap; and the Augila field of Libya, with production from fractured and weathered granite.

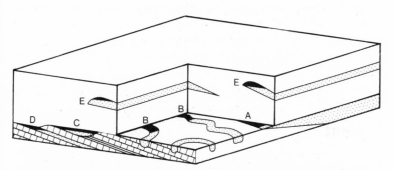

Figure 35 Stratigraphic traps in sandstones: A, supraunconformity pinchout; B, supraunconformity channels; C, supraunconformity strike-valley sand trap; D, subunconformity truncation trap; E, shoestring marine sand bars.

Stratigraphic Traps Unassociated with Unconformities

These may be best considered according to whether the reservoir is terrigenous sand or carbonates.

Sandstone Stratigraphic Traps

Sandstone stratigraphic traps are of two main classes: *channel sand bodies* and *marine sand bars*.

Channel sands are of three major types. *Fluvial channel sands* are often incised into unconformities or overbank shales and transgressed by marine shales. Examples include the Fiddler Creek, Clareton, Glenrock and Coyote fields, of Cretaceous age, in Wyoming. *Deltaic channel stratigraphic fields* are well-known from the Pennsylvanian sands of Oklahoma. Deltaic distributory channels grade into mouthbar sand bodies, which may themselves form discrete stratigraphic traps. The Pennsylvanian West Tuscola field of Texas is an example. *Submarine channels*, infilled by turbidite and grainflow sands, sometimes form stratigraphic traps. The Rosedale field of California is an example.

Marine sand bar stratigraphic traps may loosely be grouped into two types (Fig. 35): *offshore bars* and *coastal bars*.

Offshore bars, including tidal current sand bodies, occur on marine shelves some distance from the shoreline. Sand body orientation is controlled largely by tidal currents and may be unrelated to palaeoslope though is generally parallel to shoreline orientation. The Recluse field of Wyoming is an example of this type of stratigraphic trap. Coastal marine sand bars generally have a lagoonal fine-grained facies on their landward side. Their orientation is generally more predicatable, being subparallel to shoreline and palaeostrike. The Cretaceous Bell Creek field of Wyoming is of this type and is unusual in that it is a giant field (over 200 million barrels reserves).

Carbonate Stratigraphic Traps

Carbonate stratigraphic traps are far harder to locate than sandstone ones because of their eccentric diagenetic behavior.

Effective primary porosity occurs only in the Y zone of the facies model in Figure 16. Effective secondary porosity due to diagenesis may also occur, however, in the X and Z zones.

Principle carbonate stratigraphic traps are of a primary lenticular geometry (descriptively termed *buildups*), which may on detailed examination prove to be genuine reefs (Fig. 36). The Devonian reef province of Alberta illustrates the problems of this type of stratigraphic play. Reef location may be relatively easy, but it is not possible to determine if a reef is porous without drilling it, and even then it may be wet, even if it is sealed.

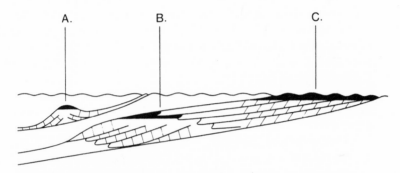

Figure 36 Potential carbonate stratigraphic traps: A, pinnacle reef play; B, shelf margin play; C, subunconformity play. Porosity in A and B may be primary or secondary, porosity in C is largely secondary solution or intercrystalline porosity.

9

The Habitat of Hydrocarbons in Sedimentary Basins

Introduction

Hydrocarbons occur in sedimentary basins. The distribution of hydrocarbons in any basin is related to the mechanics of basin formation since this controls reservoir facies, trapping mechanisms, geothermal history, and migration routes.

Mechanisms of Basin Genesis

Four major processes generate sedimentary basins (Fig. 37): (1.) Crustal compression occurs at zones of subduction. Elongated troughs of the geosynclinal suite form here. (2.) Crustal tension occurs at zones of sea floor spreading. Basins of the rift and drift suite form here. (3.) Crustal phase changes, e.g. thermal bulges, are followed by erosion and crustal collapse. Cratonic basins form here. (4.) Crustal loading occurs at continental margins where major clastic wedges depress oceanic crust.

These four processes form the basis of the genetic classification of basins shown in table 13. From their nature, basins are hard to place in a rigid classification. Table 13 shows the common transitions between the various basin types. Each of the major basins will now be described and their hydrocarbon habitats discussed.

Basins (Sensu Stricto)

These are elliptical or subcircular basins associated with continental crusts. Two subtypes may be defined.

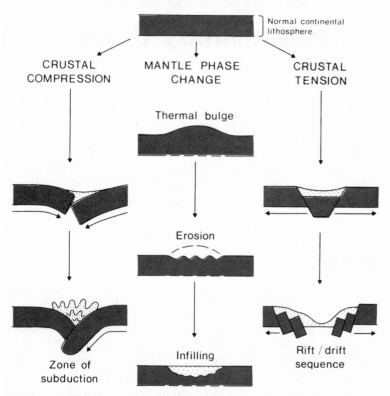

Figure 37 Basin forming mechanisms. (Alfred Fischer and Sheldon Judson, eds., Petroleum and Global Tectonics. Copyright ©1975 by Princeton University Press. Fig. 5, p. 57, adapted by permission of Princeton University Press)

TABLE 13
FOUR PROCESSES THAT FORM BASIS FOR
GENETIC CLASSIFICATION OF BASINS

Basins (sensu stricto)	Intracratonic	Cratonic suite
	Epicratonic	
Troughs	Miogeosyncline	Geosynclinal suite (associated with) subduction)
	Eugeosyncline	
	Molasse trough	
Rifts	Intramontane	Rift-drift suite (associated with sea-floor spreading)
	Intracratonic	
	Intercratonic	
Ocean margin basins		

Intracratonic Basins. Intracratonic basins are entirely surrounded by continental crusts. Examples include the Williston and Michigan basins. Though the basin floor may now be deep they are characteristically infilled by shallow water facies ranging from continental clastics to marine carbonates. Because of restricted access to the ocean evaporite and black shale facies may be well developed.

Traps in this type of basin are generally associated with regional arches. Reef traps may be present.

Intracratonic basins contain a superabundance of reservoirs but, when they lie in the middle of continental shields, may lack adequate marine source beds. Many basins of this type may have had geothermal gradients too low to have generated much oil (e.g., Murzuk and Kufra basins of Libya).

Epicratonic Basins. Epicratonic basins lie on continental crust at its margin with oceanic crust. Geometrically, these are more strictly embayments that often plunge oceanward. They contain similar facies to intracratonic basins, but have more marine deposits which may include deep water sands and muds, especially on their oceanward side. To this extent, they are transitional between intracratonic and oceanic basins.

Tectonic activity is more marked in epicratonic basins, with tensional horst and graben structures as the typical style. This may be accompanied by more volcanic activity and higher temperatures than occur in intracratonic basins.

For these reasons, epicratonic basins are often important hydrocarbon provinces with unconformity and biohermal traps associated with structural horsts. The Polignac and Sirte basins of North Africa are of this type.

Troughs

Troughs included within this grouping are linear basins associated with zones of crustal subduction. This is the realm of the geosyncline. Three main types of troughs may be included in this suite.

Eugeosynclines (axial geosynclinal troughs) are characteristically infilled by deep water sands and muds of *flysch* facies. Basic volcanics are common associates. Eugeosynclinal shales are often organic-rich and may have had considerable source potential. During orogenesis, however, high temperature and tectonic activity combine to destroy and dissipate most hydrocarbons.

Miogeosynclinal troughs, often separated from the eugeosyncline by a geanticlinal arch, are more prospective. These are also called foredeeps, lying between a craton and a fold belt. Troughs of this type contain thick sequences of shallow water sands and carbonates comparable to the fill of epicratonic basins. Foredeeps include such major oil provinces as the Arabic Gulf, the Eastern Venezuelan basins, and the North Slope of Alaska.

Factors that make these troughs such prolific producers include a balance of reservoir and source beds and a balance between the extremes of cratonic tectonic lethargy and geosynclinal excess. There are major opportunities for the migration of hydrocarbons from geosynclinal axes prior to orogenesis. These basins contain a great diversity of trap types, including regional pinchouts, reefs, salt structures, regional arches and anticlines, especially near the mountain front.

The third type of trough that is included in the geosynclinal suite is the *postorogenic molasse basin*. This group includes the Alpine marginal basins which, in the case of the Po and Rumanian basins, are significant petroleum provinces. Postorogenic molasse troughs also include barren continental facies, such as the Old and New Red Sandstones of Britain. These basins may also be considered as intermontane rifts.

The Rift-Drift Basins

Rift basins fall generically into two main types: the intermontane molasse type, previously mentioned, and the sequence of rift basins associated with axes of sea floor spreading. These last named are of three varieties, which occur in an evolutionary sequence. This begins with updoming of continental crust resulting in a system of rifts, often triradially arranged (a triple R junction). These basins are initially infilled by continental clastic sediments and volcanics.

With increasing crustal extension, the floor of the rift is depressed to sea level. At this time, the intermittent ingression of the sea is favorable for evaporite formation. With further tension and subsidence the rift becomes wholly marine, but with a restricted circulation of water. These conditions are optimal for the deposition of organic rich muds, which are potential hydrocarbon source beds. The continental crust finally separates and two half-graben basins form on either side of the new ocean floor. They are infilled by normal marine sediments. This sequence of events is exemplified by the East African Rifts, the Gulf of Suez, and Atlantic Coastal basins, such as Gabon (Fig. 38).

Basins of the rift-drift sequences are very prospective because of the vertical facies sequence of salt (capable of generating structures), organic shale (source beds), and normal marine sediments (reservoirs and seals) (Schneider, 1972). Of particular interest are the failed arms of triple-R junctions in which the rift did not drift. Here, nonetheless, the ideal facies sequence is preserved, together with high geothermal gradients and synchronous structural growth. The Central and Viking grabens of the North Sea are a case in point (Fig. 38).

Basin Analysis Review: Conclusion

The object of the preceding analysis of sedimentary basins is to show that a number of basin models may be genetically

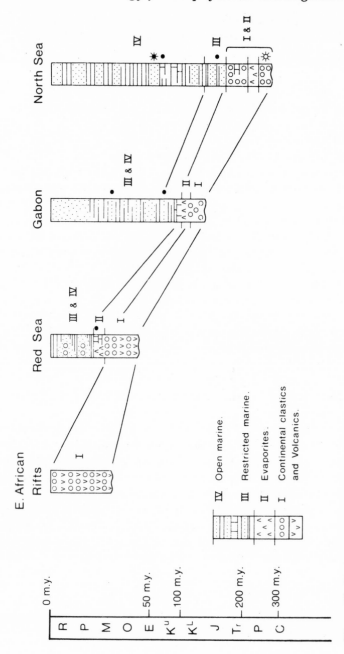

Figure 38 Facies sequences in rift basins.

defined. The petroleum potential of a basin depends on the balance between source and reservoir beds, a balance between tectonic somnolence and excess, and an optimal geothermal gradient. The hydrocarbon potential of new basins may be appraised, and old ones reappraised, by reference to these various models.

Distribution of Hydrocarbons within a Sedimentary Basin

There tends to be a regular distribution of hydrocarbons within a sedimentary basin. Within a trap, gas overlies oil; and oil overlies water. The oil column may show a gravity separation with very high gravity (low API), oil at the base, and light (high API) oil at the top. Within a sedimentary basin, however, this situation is reversed. Shallow strata are flushed by meteoric water. The shallowest hydrocarbons are heavy oils, and oil becomes lighter with increasing depth. Gas is generally encountered in the deepest part of the basin. This vertical zonation of hydrocarbons may largely be ascribed to increasing temperatures with increasing depth. (Refer back to Fig. 3.)

There is also a regional zonation of hydrocarbons within many basins. Gas occurs in the center, oil occurs in a halo, and water at the rim (Fig. 39). This zonation can be explained by Gussow's principle (Gussow, 1954). This assumes that oil and gas are generated in the center of the basin and that there are continuous permeable carrier beds that link traps from deep basin center to the shallow basin margin. The deepest traps are infilled with gas to below their spill points, forcing oil up into shallower traps until they, too, are filled to their spill points. The shallowest traps are barren of hydrocarbons, because there is not sufficient oil to fill them (Fig. 40). This is a simple elegant mechanism, but there are few basins where one can demonstrate the continuity of carrier beds that it requires.

The basin zonation can be explained purely by depth: temperature considerations, with the shallowest traps barren due to flushing by meteoric water. The spill point principle is important, however, in understanding the distribution of hydrocarbons in adjacent traps.

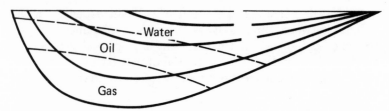

Figure 39 Diagram to illustrate the typical distribution of hydrocarbons in a sedimentary basin.

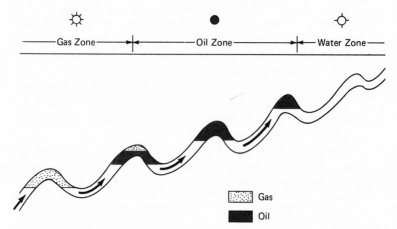

Figure 40 Diagrammatic cross section through a sedimentary basin to illustrate Gussow's principle of hydrocarbon entrapment.

Appendix

Review of the Petroleum Geology of the North Sea

Introduction

It may be useful to review the North Sea petroleum province to illustrate some of the concepts previously discussed. This account of the petroleum geology of the North Sea is divided into two parts: geological history and hydrogen habitat.

The Geological History of the North Sea

The External Framework

The North Sea and its environs contain a triangular Mesozoic-Tertiary sedimentary basin bounded by three older structural units: the *Baltic* or *Scandinavian Shield* (Precambrian) whose north-south margin defines the eastern edge of the North Sea basin, the *Caledonian Mountain Chain* (Lower Palaeozoic) an orogenic belt of igneous and metamorphic rocks with a north-east-southwest trend lying to the northwest of a line stretching from Newcastle to Bergen in the northern North Sea (a collapsed section of this zone forms the economic basement), and the *Hercynian Mountain Chain* (Upper Palaezoic) which lies to the south of a line running from north Devon, through London, to north Germany.

The Internal Framework

The North Sea basin, the margins of which are defined by the three structural features previously described, is internally dominated by two structural elements. The first element is the mid-North Sea High (RynKoping-Fyn high), with a positive axis running from northeast England to Denmark. It divides the North Sea basin into two sub-basins. The southern North Sea basin is mainly infilled by Upper Palaeozoic-Mesozoic sediments and is dominantly a gas province. The northern North Sea basin is mainly infilled by Mesozoic and Tertiary sediments and is dominantly an oil province. The second element is the Median Rift, a major rift graben that runs north to south through the North Sea, approximately coincident with the U.K. Norwegian median line. The names are the Viking Graben, the Central Graben, and the Dutch Graben (Figs. 41 and 42).

The Geological History

The North Sea is part of the crust of northwest Europe that once formed a continuous plate with Greenland. It has undergone five major structural phases that are reflected in its stratigraphy. The first two of these were intermittently tensional and compressive; the last three were exclusively tensional: the *Caledonian geosynclinal stage* [Cambrian-Silurian, culminating in orogenesis (northeast-southwest trend)], the *Variscan geosynclinal stage* [Devonian-Carboniferous, culminating in the Hercynican orogeny (west-southwest-east-northeast trend)], the *intracratonic stage* (Permian-Triassic), the

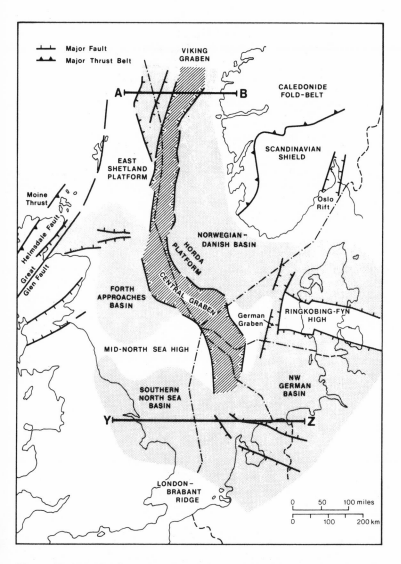

Figure 41 Map depicts major faults and basin form lines in the North Sea.

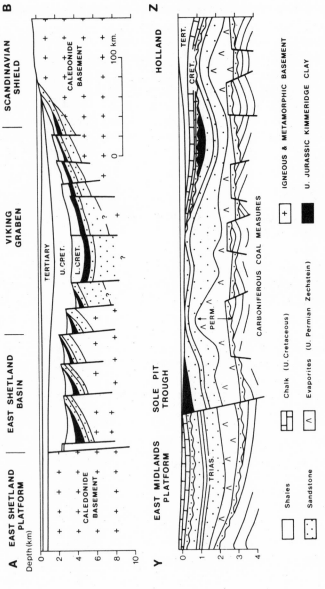

Figure 42 Cross section of the northern (upper) and southern (lower) North Sea basin. For locations see Figure 41.

taphrogenic rifting stage (Triassic-Cretaceous), and the *post-rifting, intracratonic stage* (Tertiary-Recent).

The last three of these stages relate to the opening of the North Atlantic, showing the typical sequence of sedimentary facies in a failed arm: continental clastics (Permo-Trias), evaporites (Permo-Trias), restricted marine (Jurassic), open marine (Cretaceous-Tertiary).

The Viking and Central grabens are a part of a failed rift originating from a triple-R junction to the north whose other arms opened to form the Atlantic Ocean. The occurrence of hydrocarbons in the North Sea is controlled by this structural history.

The Habitat of Hydrocarbons in the North Sea

There are four main *plays* or modes of occurrence of petroleum in the North Sea. Their characteristics are summarized in tables 14-17.

The southern North Sea basin is dominated by the Permian Rotleigende gas play. Some gas escaped upwards to be trapped in Zechstein Carbonates (Lockton) and Triassic Sandstones (Hewett). The northern North Sea basin contains three major plays, principally of oil with minor gas. Presently, the Chalk Play is restricted to southwestern Norwegian waters.

TABLE 14
HABITAT OF HYDROCARBONS IN
THE SOUTHERN NORTH SEA BASIN

Reservoir	L. Permian Rotleigende eolian sands
Seal	U. Permian Zechstein salt
Source	Gas distilled off from U. Carboniferous coal measures
Trap	Block-faulted horsts
Examples	Groningen, West Sole, Leman, Viking, Indefatigable

TABLE 15
HABITAT OF HYDROCARBONS IN THE TERTIARY DELTA

Reservoir	Submarine channel and turbidite fan sands at the foot of a major delta that built out from the Scottish coast
Seal	Tertiary shales
Source	U. Jurassic shales
Trap	Draped anticlines on deep seated horsts
Examples	The Forties and Montrose Palaeocene oil fields and the L. Eocene Frigg gas field

TABLE 16
HABITAT OF HYDROCARBONS IN THE CHALK PLAY

Reservoir	Danian and Maastrichtian chalk (coccolithic micrite)
Seal	Tertiary shales
Source	U. Jurassic shales
Trap	Gentle Zechstein salt domes
Examples	Ekofisk, Eldfisk, Tor and Dan

TABLE 17
HABITAT OF HYDROCARBONS IN THE JURASSIC PLAY

Reservoir	Rhaetic, mid and upper Jurassic paralic and (less commonly) deep-sea sands
Seal	Upper Jurassic and Lower Cretaceous shales
Source	Upper Jurassic (Kimmeridgian) shale
Trap	Tilted fault blocks truncated by unconformities, and structural noses against major graben-bounding faults
Examples	Brent, Statfjord, Thistle, Dunlin, Magnus, Hutton, Beryl, Piper, etc.

Oil also occurs in rocks of many ages that retain porosity and overlie or are over-stepped by Kimmeridgian shale: Lower Cretaceous, Zechstein, (e.g., Auk and Argyll), Carboniferous and Devonian (Buchan) (Fig. 43). Factors that have made the northern North Sea basin a major oil province include: thick sedimentary sequence, rapid sedimentation, good balance between source and reservoir rocks, deposition synchronous with structural growth, and high geothermal gradient due to crustal thinning.

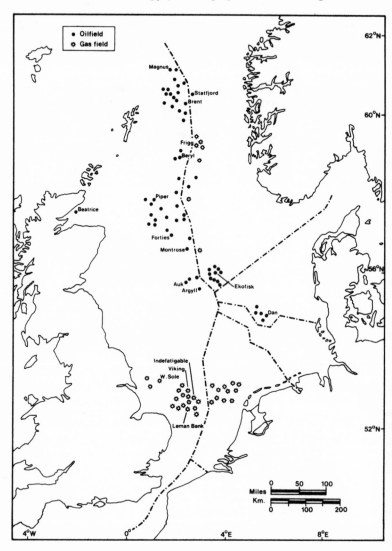

Figure 43 Map depicts oil and gas fields in the North Sea.

References

Fisher, A.G., and S. Judson, 1975, Petroleum and global tectonics:
    Princeton, Princeton Univ. Press, 322 p.

Gussow, W.C., 1954, Differential entrapment of oil and gas: a fundamental
    principle: AAPG Bull., v. 38, p. 816-852.

Halbouty, M.T., et al., 1970, Worlds giant oil and gas fields: AAPG Mem.
    14, p. 502-556.

Irwin, M.L., 1965, General theory of epeiric clear water sedimentation:
    AAPG Bull., v. 49, p. 445-459.

Kent, P.E., 1975, Review of North Sea basin developments: J. Geol. Soc.
    Lond., v. 131, p. 435-468.

Lees, G.M., 1953, Persia: science of petroleum, volume 6: New York City,
    Oxford University Press Inc., p. 73-82.

Levorsen, A.I., 1967, Geology of petroleum: second edition, San Francisco,
    W.H. Freeman and Company, 724 p.

Moody, J.D., J.W. Mooney, and J. Spivak, 1970, Giant oil fields of North
    America: AAPG Mem. 14, p. 8-18.

Rittenhouse, G., 1972, Stratigraphic trap classification: AAPG Mem. 16,
    p. 14-28.

Schneider, E.D., 1972, Sedimentary evolution of rifted continental margins:
    Geol. Soc. America Mem. 132, 109-118.

Selley, R.C., 1976, The habitat of North Sea oil: Proc. Geol. Ass. Lond.,
    v. 87, p. 359-388.

_____, 1976, Subsurface environmental analysis of North Sea sediments: AAPG Bull., v. 60, p. 184-195.

_____, 1978, Ancient sedimentary environments, second edition: Ithaca, New York, Cornell University Press, 287 p.

Terry, C.E., and J.J. Williams, 1969, The exploration for petroleum in Europe and North Africa: London, Institute of Petroleum.

Ziegler, P.A., 1975, Geologic evolution of the North Sea and its tectonic framework: AAPG Bull., v. 59, p. 1073-1097.

Index

Asphalt, 2

Boreholes, 10, 15-23
Buildups, 67

Cap rock(s), 3
  evaporites as, 51, 54
Carbonate(s), 41-49
  environments, 43-46
  grain type, 42, 43
  mineralogy, 42-43
  stratigraphic traps, 66-67
Catagenesis, 9-11
Channel sand bodies, 66
Clay(s)
  minerals, 8-9
  organic matter in, 8-9
  of source rock, 3
Combination traps, causes of, 58
Combination well logs, 23
Condensates as hydrocarbons, 2
Crude oil, 2, 8

Darcy's Law, 32-34
Density log, 18-19
Diagenesis, 9
Dipmeter log(s), 21-23, 46

Epicratonic basins, 71-72
Eugeosynclines, 72
Evaporating dish mechanism, 53
Evaporites, 51-56
  as cap rocks, 51, 54
  formation of, 53
  occurrence of, 52-53
  as source rocks, 56
  as traps, 54-55

Facies analysis, integrated, 23
Facies analysis, subsurface, 23
  for carbonates, 42
  of reefs, 46
  of reservoir rock, 37
  for sandstones, 42
Facies defined, 37
Facies model, carbonate, 43
Formation evaluation defined, 15

Gamma logs, 18, 23, 24
Gas, 2. See also Oil and gas.
Geophysical well logs, 15-29
  methods for, 17-23
Geosynclinal suite, 72-73
Giant field defined, 3
Grainsize profiles, 23-24
Grain types, carbonate, 43

Hydrocarbon(s), 1-5
  accumulation, parameters for, 3
  distribution of, 75-76
  migration of, 11-13
  in the North Sea, 81-84
  properties of, 1-2
  in sedimentary basins, 69-76
  traps, 57-62
Hydrodynamic traps, 57-58, 59

Intracratonic basins, 71

Kerogen
  and catagenesis, 9-10
  as hydrocarbon, 1-2

Laevorotation defined, 8
Linear basins. See Troughs

87

Logs. *See* Geophysical well logs

Magic five, the, 1
Marine sand bars, 66
Metagenesis, 9
Migration of hydrocarbons, 11–13
Mudlogging defined, 15

Neutron log, 18
North Sea, the, geology of, 77–84

Oil and gas. *See also* Crude oil; Gas
  fields, 3–5
  traps for, 3
Organic matter
  destruction rate of, 8
  generation rate of, 7–9
  preservation of, 8–9

Palaeothermometers, 10
Permeability, 31, 32–34
Porosity
  location of, 42
  logs for, 19
  of reservoir rocks, 31–32
Primary migration, 11–13

Reefs, 45–46
  as stratigraphic traps, 67
Reservoir rocks, 31–39
  carbonates as, 41–49
  environment of, 37
  and hydrocarbon accumulation, 3
  permeability of, 31, 32–34
  porosity of, 31–32
Resistivity log, 17

Sand, 35–37
Sandstone(s)
  classification of, 34–35
  Cretaceous, and oil field, 3
  defined, 34
  reservoir rocks, 34–39
  stratigraphic traps, 66
  subsurface facies analysis for,
    42
Secondary migration, 12
Sedimentary basin(s)
  classification of, 70–71
  described, 70–72
  distribution in, 75
  generation of, 69–70
  hydrocarbons in, 69–76
  in the North Sea, 77–78
  potential of, 75
  rift-drift, 73

Sedimentary environments, 37
Sensu Stricto. *See* Sedimentary
  basin(s)
Sonic log, 19–21
Source rock(s)
  evaporites as, 51, 56
  hydrocarbon, 7–13
  and hydrocarbon accumulation, 3
  kerogen in, 2
  measuring temperature of, 10–11
Sour gas, 2
Spontaneous potential log (S.P.),
  17, 23, 24
Stratigraphic traps, 63–67
  causes of, 57
  defined, 63
  locating, 63–64
  types of, 64–67
Stratigraphy, North Sea, 78–81
Structural traps, 57, 58–59
Structure of the North Sea, 78–81
Subsurface facies analysis, *See*
  Facies analysis, subsurface
Subunconformity stratigraphic
  traps, 65
Supraunconformity stratigraphic
  traps, 64–65
Sweet gas, 2
Swelling clays, 9

Tadpole plot, 21
Temperature
  and boreholes, 10
  and catagenesis, 10–11
  and organic matter, 7
  and hydrocarbon accumulation, 3
Terrigenous rocks, 41
Terrigenous sands, 34
Traction currents, 35–37
Traps, 3
  evaporites as, 51, 54–55
  hydrocarbon, 57–62
  hydrodynamic, 57–58, 59
  for oil and gas, 3
  stratigraphic (*See* Stratigraphic
    traps)
  structural, 57, 58–59
Troughs, types of, 72–73
Turbidity currents, 35–37

Unconformity-associated
  stratigraphic traps, 64

Well logs. *See* Geophysical well
  logs
Wireline well logging, 15